HITLER'S WAR MACHINE

THE PANTHER V IN COMBAT

GUDERIAN'S PROBLEM CHILD

BOB CARRUTHERS

C✛DA
BOOKS LTD
www.codabooks.com

This edition published in Great Britain in 2012 by
Coda Books Ltd, The Barn, Cutlers Farm Business Centre, Edstone,
Wootton Wawen, Henley in Arden, Warwickshire, B95 6DJ
www.codabooks.com

BOOKS LTD

Copyright © 2012 Coda Books Ltd.

A CIP catalogue record for this book is available from the British Library.

ISBN: 978-1-78158-065-3

CONTENTS

- C H A P T E R   1 -
GUDERIAN'S PROBLEM CHILD

FROM LATE 1942 onwards the German engineers employed in the hard pressed armaments industry produced a remarkable range of armoured fighting vehicles. They were driven by the desperate demands of an insatiable front line which was showing the first warning signs that it might ultimately roll back and consume their homeland in a red tide. The race against time was remorseless, and this ominous situation was compounded by Adolf Hitler who harboured unrealistic expectations that new and improved tank designs were somehow capable of turning back the surge of the Red Army. The weight of Hitler's expectancy and his unreasonable deadlines placed the engineers and manufacturers under extreme pressure to design, develop and supply new and unproven battlefield technology as quickly as possible. The war in the East was a demanding and remorseless taskmaster which consumed every

The Panzer Mark V - The Panther.

4

new offering as soon as it was ready for action, the price of failure was unthinkable and, not surprisingly, this unsettling combination of concerns actually drove the German armaments industry on to some remarkable achievements. Chief among these was the development and deployment of the Panzer Mark V - The Panther.

Among the many excesses which the Nazi regime condoned was its willingness to embrace the concept of slave labour. The German armaments industry workforce, both willing and enslaved, worked ceaselessly in gloomy war ravaged factories to design, develop and produce an astonishing variety of highly effective armoured fighting vehicles which appeared on the battlefield in an incredibly short period of time. It has often been said that the German armaments industry placed the best possible weapons in the worst possible hands; that is certainly true of the Panther.

Under peacetime conditions a new fighting vehicle would generally be designed, built and tested over a period of three to five years. Between 1941 and 1945 however some very successful designs for armoured fighting vehicles were produced in just 12 months. As war progressed the Red Army received huge volumes of increasingly sophisticated vehicles and weapons and as a result it became imperative that new German vehicles should be brought into action as quickly as possible. Many of the most successful German machines were adaptations of existing vehicles which were modified to produce specialist tank destroyers such as the *Jagdpanzer* IV and the Nashorn. Other armoured fighting vehicles such as the Tiger and the Panther, were completely designed and built from scratch, these designs were more innovative and, in most respects, more effective, but the hasty development process meant that significant teething problems often remained unsolved. In the case of the Panther the evidence of this hurriedness was evident for all to see. On his way up to the front lines prior to the battle of Kursk SS Panzer Grenadier Hofstetter recalled seeing the new Panther for the first time.

Heinz Wilhelm Guderian was a German general during World War II. He was a pioneer in the development of armored warfare, and was the leading proponent of tanks and mechanization in the Wehrmacht. Germany's panzer forces were raised and organized under his direction as Chief of Mobile Forces. During the war, he was a highly successful commander of panzer forces in several campaigns, became Inspector-General of Armored Troops, rose to the rank of Generaloberst, and was Chief of the General Staff of the Heer in the last year of the war.

"As we passed the unfamiliar column of Panzers, it was soon obvious that there had been a serious problem with one machine in particular that was reduced to a burnt out wreck with no sign of any enemy activity. We later learnt that this was the Panther – Guderian's problem child!"

The omens pointed towards dismal failure, but against heavy odds, the Panther gradually gained a fearsome reputation and eventually produced a legacy which shaped the face of tank design in the post war era. The Panther's excellent combination of firepower and mobility produced a fighting machine which has frequently been hailed as one of the best tank designs of World War II. It has been estimated that every Panther deployed accounted for, on average, five allied tanks and as many as nine Russian tanks.

The statistics are unproven and in reality it may well be the case that the Panther was actually a very expensive failure which drew much needed resources away from the real requirements of the *Panzertruppen*. At the time it was strongly argued, by Guderian and others, that what the hard pressed front line troops really needed was to a high volume of reliable main battle tanks in the shape of the Mark IV F2, a machine which could at least attain parity with the T-34. There remains the strong argument that the decision to develop the Panther was a wrong option, especially as the problems caused by the poor quality components used in the final drive were never overcome. This serious flaw made the Panther highly susceptible to breakdowns which were so frequent as to be almost a certainty. These catastrophic mechanical failures were particularly common during the long road marches which became increasingly numerous as the war progressed and the Panthers had to be rushed from place to place. Many of the repairs required, particularly those on the final drive, were very difficult and frequently could not be managed at divisional workshops and required the vehicle being returned to army depots far behind the lines. On the retreat it was not always possible to

Minister of Armaments and War Production for the Third Reich, Albert Speer, inspects a Panther tank. The vehicle has already received a factory coating of Zimmerit anti-magnetic paste.

recover the broken down vehicles and this led to many, otherwise salvageable, Panthers having to be destroyed.

The other principal disadvantage of the Panther lay in its comparatively weak side armour which made it highly vulnerable to attack from any direction other than head on. As a result of the poor ammunition stowage design, the tank was also highly susceptible to "brewing up" when hit. Taken together these negative aspects of the Panther are the main reasons why the Panther has not attained the legendary status of the Tiger I which was much loved by its crews.

Like the Tiger, the development of the Panther resulted from the Wehrmacht's unpleasant surprise encounter with the Soviet T-34 during Operation Barbarossa, the German invasion of Russia, in June 1941. During the first weeks of Barbarossa, the men of the *Panzertruppen* repeatedly encountered the T-34/76 medium tank. Although in short supply, the T-34 made a quick and lasting impression on the German armoured forces who were shocked to be confronted with this formidable

The original T-34 Model 1940 - recognizable by the low-slung barrel of the L-11 gun below a bulge in the mantlet housing its recoil mechanism. This pre-production A-34 prototype has a complex single-piece hull front.

9

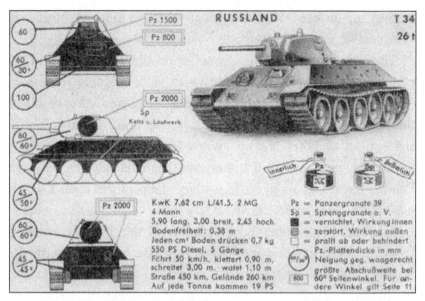

A page from the Pantherfibel the crew manual published in 1944 demonstrating graphically how to combat the T-34 visually illustrating the weak spots (in black) as aiming points.

vehicle with its near perfect combination of speed and mobility, rugged reliability, sloped armour protection and firepower. As a result of numerous adverse encounters with the T-34, especially the battering sustained by the 4th Panzer division at Mtsensk on 4th October 1941, Colonel General Heinz Guderian, leading *Panzergruppe* 2 in Army Group Centre, requested the establishment of a Commission of Enquiry into the relative strengths of the tank armies on the Eastern Front.

Although Guderian suggested simply copying the T-34, this proposal was rejected and the report of the enquiry instead recommended that the main attributes of the T-34 be incorporated into a new German built machine. The main points which were desirable in the new design the T-34 were its excellent main armament which was capable of firing both high velocity anti-tank rounds and a reasonably effective high explosive shell , well sloped armour and a highly effective suspension design with wide tracks which gave good cross country mobility.

10

The outcome of Hitler's intervention in the debate was the decision to produce a brand new medium tank – the Panther. Hitler however demanded a crash building programme and as soon as the new machine was off the drawing board and into production, the Panther underwent a complex and difficult development cycle which included overcoming problems with the vehicle's transmission, steering, main gun, turret and fuel pump. Despite having to contend with these and a host of other issues the first 200 Panthers were nonetheless readied for participation in the Wehrmacht's 1943 summer offensive in the East. The Panther then saw action from mid-1943 to the end of the European war in 1945. It was intended as a counter to the T-34, and to replace the Panzer III and Panzer IV. However a remorseless allied bombing campaign meant that production did not reach the necessary levels and Panther formations served alongside those equipped with the Mark IV and the heavier Tiger tanks until the end of the war.

CHAPTER 2

THE PANTHER MANUAL:
THE PANTHERFIBEL

THE CREW of the Panther comprised five members: a driver, a radio operator (who also operated the bow machine gun), a gunner who aimed and fired the main gun and co-axial machine gun , a gun loader, and a commander.

The crews selected for duty on the Panther were selected among the very best that could be found. Due to the high number of teething issues the Panther required extremely sympathetic handling by knowledgeable crews, but by 1943 the process of identifying and training crews was becoming increasingly difficult. Maintaining and fighting the Panther demanded a great deal of specialist knowledge, both theoretical and practical, which required long hours of class room study as well as, ideally, months familiarisation and training on the Panther itself. However, the

The introductory page from the Pantherfibel which offered the crew the opportunity to learn with ease.

The five man crew of the Panther tank in Italy, August 1944.

declining war situation and a shortage of training machines dictated that the time and resources which would normally be allocated to crew training were seriously curtailed.

One successful element of the training programme was the introduction *Pantherfibel* an illustrated crew manual which followed the style of the highly successful Tiger I crew training manual, the *Tigerfibel*. Surviving copies of the The *Pantherfibel* provide a fascinating primary source insight into the world of the men of the *Panzertruppen* who crewed the Panther and extensive use has been made of the *Pantherfibel* throughout this book.

The man responsible for the evolution of the *Pantherfibel* was *Oberstleutnant* Hans Christern, head of training for the Inspectorate of the *Panzerwaffe* based at Paderborn. Christern was an experienced tank commander who could provide proof to his own practical experience by dint of his the possession of the Knight's Cross awarded to him for bravery in the field.

With the introduction of the Tiger I (Ausf H) in late 1942 Christern found himself faced with the need to rapidly instruct crews in the operation of a very different type of vehicle. Like the Panther this tank had to be handled very differently on the battlefield like the Panther it needed far more care and attention than any other machine so far delivered to the *Panzerwaffe*. The *Tigerfibel* therefore dealt with the same set of issues which were encountered with the later introduction of the Panther.

Faced with a rapidly declining war situation everything needed to be done in a hurry. Christern therefore decided it would help to move matters along if he were to replace the usual dusty tank instruction manual with a special training booklet for Tiger I students which was simple yet memorable. The end result was certainly a success on both counts. The simplistic but effective style recalled a children's school book. It was therefore given the name *Tigerfibel*, which means Tiger primer. This booklet was assigned the official publication number of D656/27.

The Red Army menace is characterised as a ravenous bird of prey in this page from the Pantherfibel which cautions the men of the Panzertruppen to be constantly vigilant.

A Panther tank commander surveys the surrouding area on the eastern front in the summer of 1944.

The task of actually writing the *Tigerfibel* was assigned to *Leutnant* Josef von Glatter-Goetz. Glatter-Goetz took the assignment to heart and gave serious consideration to the need to impart such a large amount of information quickly and make it stick in the minds of bored young tank men. He therefore developed the idea of writing a humorous and highly risqué manual that would hold fast in the memories of the young men training on the Tiger I. To do this he used humorous and risque cartoon illustrations along with slang and the everyday situations which it was hoped the target audience would identify with.

The illustrations in the *Tigerfibel* were completed by two serving soldiers named *Obergrenadier* Gessinger and *Unteroffizier* Wagner. This wide range of images included the usual technical drawings and photographs supplemeted by a range of cartoons. Wherever possible the cartoons featured an attractive and curvaceous blonde named Elvira, depicted naked as often as possible and somewhat predictably was the romantic target for the affections of a Tiger crewman who gets the girl in the end.

The radio duty of the Commander from the Pantherfibel.

The *Tigerfibel* also contains some short verses and rhyming couplets which do not lend themselves readily to an exact translation from German and English.

Both The *Tigerfibel* and the later *Pantherfibel* (published on 1st July 1944) included vital information concerning basic maintenance requirements and peculiarities, and in addition covered a wide range of additional subjects. There was important advice on gunnery and ammunition drill as well as a comprehensive run down the type of enemy targets likely to be encountered. In addition came advice on driving techniques, winter conditions, fuel conservation, how to deal with enemy infantry at close quarters, anti-tank mines, target spotting and a host of additional information.

Although it was quite unconventional when compared to any other manual hitherto produced and was somewhat racy by straight laced Third Reich standards, the *Tigerfibel* was actually authorized by Guderian himself and it proved to be very effective training aid. The *Pantherfibel*, which followed a year later was again personally authorised by Guderian, however it is the product of a far more sober environment and presents a much less saucy publication which reined in the gratuitous cartoon nudity in preference for some rather twee rhyming couplets which obviously don't have the same instant appeal to the common soldier. The *Pantherfibel* nonetheless featured good black and white photos and diagrammatic representations of the various Allied tanks which the Panther crew could be expected to encounter in the field. Another particularly interesting feature is the graphic cloverleaf demonstration of the range the Panther could be penetrated by or itself penetrate enemy tanks such as the Sherman M4 or T-34 or KV1.

In addition to the advice on fighting and maintaining the machine the *Pantherfibel* also affords a fascinating insight into the deteriorating supply situation in the form of exhortations to conserve ammunition and to overrun targets rather than use precious shells. Both the *Tigerfibel* and the *Pantherfibel* are also noteworthy for the fact that, despite Guderian's strong Nazi sympathies, no Nazi iconography appears anywhere in either the *Tigerfibel* or the *Pantherfibel*.

- C H A P T E R   3 -
THE RE-BUILDING PROGRAMME

THE PANTHER was to all intents a prototype, but Hitler was intent on seeing the new tank in action by the middle of 1943. However by late 1942, with production now in progress it was all too obvious that the machines emerging from the production lines were far from perfect. A rebuilding and final proofing programme was therefore introduced in order to try and deal with the remaining teething troubles. Under very difficult circumstances the DEMAG factory entrusted with the rebuild programme managed to deliver 200, ostensibly combat ready, Panthers to the Eastern Front in time to make an operational debut in Operation *Zitadelle*, otherwise known as the Battle of Kursk. The last great German offensive in the East began on 5th July 1943 and the two battalions of Panthers involved were split across the 51st and 52nd Panzer Battalions, which were attached to the *Grossdeutschland Panzergrenadier* Division on the southern flank of the Kursk salient. Inevitably, the continuing mechanical and design flaws and the limited time available for training had a disastrous effect. There was simply no available time to properly

The Pantherfibel emphasised the role of the hard pressed home front in producing the essential parts of the Panther.

train the crews and this, when combined with the mechanical problems, severely hampered the Panthers' contribution to *Zitadelle*. Despite the fact that the Panthers on their first combat foray were credited with 267 enemy tanks destroyed, it was at Kursk, and for good reason, that the reputation of the Panther as *Guderian's problem child* took root. Germany simply did not have the resources to be able to loose tanks at this ratio and this sobering fact was obvious to all.

The rushed emergence of the Panther into action at Kursk not only compromised frontline performance, it also left a large number of Panther wrecks on the battlefield. The ability of the Russians to study captured machines meant that the cat was out of the bag and a series of counter measures designed to defeat the Panther in action were soon being implemented. The following report on the new German Panther tank, based on Russian intelligence sources, appeared in the US intelligence manual *Tactical and Technical Trends*, on November 4, 1943. As the Panther tank was first deployed on the Russian front initial US intelligence on the Panther tank was based entirely on Russian sources.

THE CONTEMPORARY VIEW 1
THE PZ-KW 5 (PANTHER) TANK

The German tank series 1 to 6 has now been filled in with the long-missing PzKw 5 (Panther) a fast, heavy, well-armored vehicle mounting a long 75-mm gun. It appears to be an intermediate type between the 22-ton PzKw 4 and the PzKw 6 (Tiger) tank. The Panther has a speed of about thirty-one miles per hour. It approximates (corresponds roughly to) our General Sherman, a tank which evoked complimentary comment in the Nazi press.

The following is a description of the tank: (It should be noted that practically all data contained in this report come from Russian sources).

Weight	45 tons
Crew	5
Armament	75-mm (2.95 in) gun, long barrel, (1943) 1 machine gun, MG-42, 7.92-mm
Ammunition	75 rounds (AP & HE)
Motor	Maybach, gasoline, 640 hp in rear of tank The gas tanks are located on either side of motor
Cooling System	Water
Ignition	Magneto
Armor	Front of turret and cannon shield 100 mm (3.94 in) Upper front plate 85 mm (3.45 in) 57° inclination Lower front plate 75 mm (2.95 in) 53° inclination Side and rear plate 45 mm (1.78 in) Top of turret & tank and bottom of tank 17 mm (.67 in)
Dimensions:	
Width	11 ft 8 in (same as the PzKw 6)
Length	22 ft 8 in (1 1/2 ft longer than the PzKw 6)
Clearance	1 ft 8 in (10 cm)(3.9 in) more than the PzKw 6)
Caterpillar Section	Drive sprockets at front; rear idlers; 8 double rubber-tired bogie wheels 850 mm (33.46 in) in diameter on either side; torsion suspension system; hydraulic shock absorbers located inside tank; metal caterpillar tread 660 mm (25.62 in) wide
Maximum Speed	50 km hr. (approx. 31 mph)
Range	170 km (approx. 105 miles)

The 75-mm gun is probably the new Pak. 41 AT gun with a muzzle velocity of 4,000 foot-seconds. The estimated armour penetration at 547 yards is 4.72 inches, and the life of the barrel from 500 to 600 rounds. The gun has direct sights to 1,500 meters or 1,640 yards. The 75-mm has an overall length of 18 feet 2 inches.

The Panther can also be easily converted for fording deep streams by attaching a flexible tube with float to the air intake. There is a special fitting in the top rear of the tank for attaching this tube.

Although provided with smaller armour and armament than the 6, the Panther has the same motor, thus giving it higher speed and manoeuvrability. This tank is also provided with light armour plate (not shown in the sketch) 4 to 6 millimetres thick along the side just above the suspension wheels and the inclined side armour plate.

A Panther left destroyed in the aftermath of the Battle of Kursk.

Panther tanks are organized into separate tank battalions similar to the Tiger tanks. Many of these tanks have been used by the Germans during the July and August battles. The Russians state that this tank, although more manoeuvrable, is much easier to knock out than the PzKw 6. Fire from all types of rifles and machine guns directed against the peep holes, periscopes and the base of the turret and gun shield will blind or jam the parts. High-explosives and armour-piercing shells of 54-mm (2.12 in) calibre or higher, at 800 meters (875 yds) or less, are effective against the turret. Large calibre artillery and self-propelled cannon can put the Panther out of action at ordinary distances for effective fire. The inclined and vertical plates can be pierced by armour-piercing shells of 45 mm calibre or higher. Incendiary armour-piercing shells are especially effective against the gasoline tanks and the ammunition located just in the rear of the driver.

The additional 4 to 6 mm armour plate above the suspension wheels is provided to reduce the penetration of hollow-charge shells but the Russians state that it is not effective. Antitank grenades, antitank mines and "Molotov cocktails" are effective against the weak bottom and top plates and the cooling and ventilating openings on the top of the tank just above the motor.

This tank is standard but the quantity and rate of production is not known.

The negative lessons of Kursk were many, but those improvements which could be made were quickly absorbed and modifications were adapted into the production lines. Improvements included stronger, lower-profile commander cupolas, rain guards on the gun mantlet, Zimmerit anti-magnetic

A close view of Close view of Zimmerit on the turret of a Panther in Italy in 1944. The coating was created by the German company Chemische Werke Zimmer AG.

mine paste and, on the later Ausf. G, a simplified and strengthened hull. Given the production difficulties and the complex internal politics of German weapons manufacture, the Panther tank was inevitably a compromise of various requirements. It shared essentially the same engine as the Tiger I tank, it had better frontal armour, better gun penetration, was lighter overall, faster, and could handle rough terrain better than the Tigers.

The Achilles heel of the Panther was resultant trade off in the provision of weaker side armour which made the tank highly vulnerable to attack from any direction other than head on. Setting aside this glaring weakness the Panther still proved to be a fearsome adversary in open country especially in long range gunnery duels, but the Panther was extremely vulnerable in close-quarter combat. It should also be noted that the 75 mm gun fired a smaller shell than the Tiger's 88 mm gun, providing significantly less high explosive firepower against infantry and soft skinned targets.

The Panther was however far cheaper to produce than the Tiger tanks, and only slightly more expensive than the Panzer IV. The reason for this was unexpected outcome was that its production run coincided with the Reich Ministry of Armament and War Production's improved efforts to increase the efficient production of war materials. Nonetheless it cannot be stressed enough that key elements of the Panther design, such as its armour, transmission and final drive, were compromised and reductions in quality were made specifically to improve production rates and address Germany's requirement for numbers on the battlefield. Particularly with regard to the final drive, this was to prove a false economy.

Ironically other expensive and over engineered elements such as the Panther's complex suspension system remained. The net result of the various compromises was that some progress towards the achievement of the ambitious production goals was made and with 6000 machines being produced between 1943 and

The main armament of the Panther - a 75 mm KwK 42 (L/70), 1944.

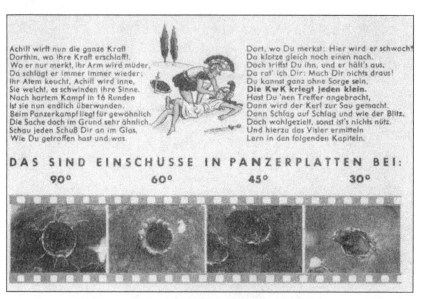

An important lesson from the Pantherfibel was the effect of the angle at which a shot struck the defensive plate.

1945 Panther tank production ran at a far higher rate than was possible for the Tiger tanks which saw only 1800 machines of both types produced between 1942 and 1945. It would appear on best estimates available however that the production run of 6000 Panthers was achieved at a very high opportunity cost. In all probability the Panther was created at the expense of some 20,000 Mark IV tanks which could otherwise have been built.

The main gun on the Panther was the 7.5 cm Rheinmetall-Borsig KwK 42 (L/70) with semi-automatic shell ejection and a supply of 79 rounds (82 on the Ausf. G). The main gun used three different types of ammunition: APCBC-HE (Pzgr. 39/42), HE (Sprgr. 42) and APCR (Pzgr. 40/42), the last of which was usually in short supply. While it was of only average calibre for its time, the Panther's gun was in fact one of the most powerful tank guns of World War II, this was due to the large propellant charge and the long barrel, which gave it a very high muzzle velocity and excellent armour-piercing qualities. The flat trajectory also made hitting targets much easier, since accuracy was less sensitive to range. The Panther's 75 mm gun had more penetrating power than the main gun of the Tiger I heavy tank, the 8.8 cm KwK 36 L/56, although the larger 88 mm projectile might inflict more damage if it did penetrate.

The tank typically had two MG 34 machine guns of a specific version designed for use in armoured combat vehicles featuring an armoured barrel sleeve. An MG 34 machine gun was located co-axially with the main gun on the gun mantlet; an identical MG 34 was located on the glacis plate and fired by the radio operator. Initial Ausf. D and early Ausf. A models used a "letterbox" flap opening, through which the machine gun was fired. In later Ausf A and all Ausf G models (starting in late November-early December 1943), a ball mount in the glacis plate with a K.Z.F.2 machine gun sight was installed for the hull machine gun.

The front of the turret was a curved 100 mm thick cast armour mantlet. Its transverse-cylindrical shape meant that it was more likely to deflect shells, but the lower section created a shot trap.

Panther with regular mantlet.

If a non-penetrating hit bounced downwards off its lower section, it could penetrate the thin forward hull roof armour, and plunge down into the front hull compartment. Penetrations of this nature could have catastrophic results, since the compartment housed the driver and radio operator sitting along both sides of the massive gearbox and steering unit; more importantly, four magazines containing main gun ammunition were located between the driver/radio operator seats and the turret, directly underneath the gun mantlet when the turret was facing forward.

From September 1944, a slightly redesigned mantlet with a flattened and much thicker lower "chin" design started to be fitted to Panther Ausf. G models, the chin being intended to prevent such deflections. Conversion to the "chin" design was gradual, and Panthers continued to be produced to the end of the war with the rounded gun mantlet.

In most cases the Panther's gun mantlet could not be penetrated by the M4 Sherman's 75 mm gun, the T-34s 76.2 mm gun, or the T-34-85s 85 mm gun. But it could be penetrated by well-aimed shots at 100 m by the 76mm M1A1 gun used on certain models of the M4, at 500 m by the Soviet A-19 122 mm gun on the IS-2 and at over 2500 yards (2286 m) by the British Ordnance QF 17 pounder using APDS ammunition. The side turret armour of 45 mm (1.8 in) was vulnerable to penetration at long range by almost all Allied tank guns, including the M4's 75 mm gun which could penetrate it at 1,500 m (0.93 mi). These were the main reasons for continued work on a redesigned Panther turret, the Schmalturm.

The Ausf. A model introduced a new cast armour commander's cupola, replacing the more difficult to manufacture forged cupola. It featured a steel hoop to which a third MG 34 or either the coaxial or the bow machine gun could be mounted for use in the anti-aircraft role, though it was rare for this to be used in actual combat situations.

The first Panthers (Ausf. D) had a hydraulic motor that could traverse the turret at a maximum rate of one complete revolution

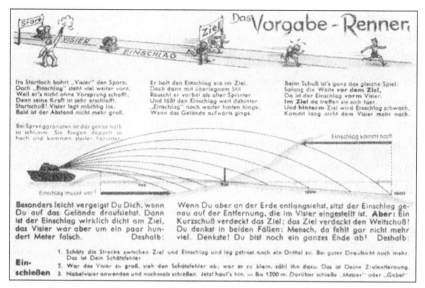

Advice on gunnery from the Pantherfibel.

in one minute, independent of engine speed. This slow speed was improved in the Ausf. A model with a hydraulic traverse that varied with engine speed; one full turn taking 46 seconds at an engine speed of 1,000 rpm but only 15 seconds if the engine was running at 3,000 rpm. This arrangement was a slight weakness, as traversing the Panther's turret rapidly onto a target required close coordination between the gunner and driver who had to run the engine to maximum speed. By comparison, the turret of the M4 Sherman turret traversed at up to 360 degrees in 15 seconds and was independent of engine speed, which gave it an advantage over the Panther in close-quarters combat. As usual for tanks of the period, a hand traverse wheel was provided for the Panther gunner to make fine adjustment of his aim.

Ammunition storage for the main gun was a weak point. All the ammunition for the main armament was stored in the hull, with a significant amount stored in the sponsons. In the Ausf. D and A models, 18 rounds were stored next to the turret on each side, for a total of 36 rounds. In the Ausf. G, which had deeper sponsons, 24 rounds were stored on each side of the turret, for a total of 48 rounds. In all models, 4 rounds were also stored in the left sponson between the driver and the turret. An additional 36 rounds were stored inside the hull of the Ausf. D and A models - 27 in the forward hull compartment directly underneath the mantlet. In the Ausf. G, the hull ammunition storage was reduced to 27 rounds total, with 18 rounds in the forward hull compartment. For all models, 3 rounds were kept under the turntable of the turret. The thin side armour could be penetrated at combat ranges by many Allied tank guns, and this meant that the Panther was vulnerable to catastrophic ammunition fires ("brewing up") if hit from the sides.

The loader was stationed in the right side of the turret. With the turret facing forward, he had access only to the right sponson and hull ammunition, and so these served as the main ready-ammunition bins.

Thanks to the Kwk 42 L/70 main gun The Panther offered a superb performance at longer ranges with excellent accuracy and a very high muzzle velocity which posed an extreme danger for

A Panther with a destroyed engine bay at the roadside in Normandy.

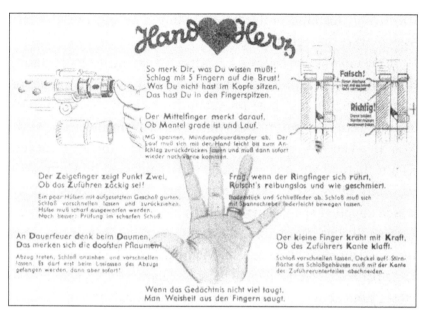

Advice on operating the machine gun in the Panther from the Pantherfibel.

every enemy tank. The tried and tested Panzer IV Ausf. G came on stream in April 1943 and although it was equipped with the less powerful KwK 40 L/48 main gun it nonetheless offered a similar battlefield performance to the Panther at shorter ranges and most importantly did not suffer from the appalling final drive issues which became the single major cause of breakdowns of the Panther tank, and which remained a problem throughout its service life.

The Panther was one of many German weapon systems with which Hitler became fixated. He placed a great deal of faith in his own assumption that the Panther could deliver a major contribution towards turning the course of the war in Russia. This high level of personal expectation, and the resulting pressure from the Fürher, led to the vehicles being rushed through the design process and into combat long before they were ready. The Panthers duly arrived on the battlefield in 1943 at a crucial phase in World War II for Germany and were rushed into combat at Kursk with a glaringly inefficient final drive system and before its obvious teething problems, including a porous fuel delivery system were corrected.

In the months that followed, to a limited extent, the most glaring difficulties were overcome. The Panther tank thereafter fought on outnumbered on the most important fronts as the German army steadily retreated before the Allies for the remainder of World War II. The faint possibility of the Panther proving a success as a battlefield weapon was drastically hampered by Germany's generally declining position in the war. The long logistical tail which supplied spare parts gradually dried up and major repairs became all but impossible to effect. With the loss of air cover more and more Panthers became victims of allied interdiction. The large poorly protected engine deck was particularly vulnerable to attack from above, and it has been estimated that 70% of Panthers were destroyed by aerial attack. As the war wore on towards a conclusion the desperate fuel situation led to many broken down Panthers, even those awaiting only minor repairs being destroyed. The pressure on training personnel and facilities allied to the declining quality of tank crews meant that the Panther, a tank

Panther wreckage Nortmandy at Kursk.

32

which required the very best crews was often handled by novices and therefore faced a massive range of obstacles which could not be overcome.

It is a mark of the fighting qualities of the Panther that, despite all the factors ranged against it, this late introduction to the war, with its favourable combination of fire power and heavy frontal armour still drew accolades from the allies who fought against it. As a result of its high kill ratio in combat the Panther soon became feared and respected by the Allies, and regardless of its many short comings has become known to posterity one of the best all-round tanks of the war.

- CHAPTER 4 -
THE DEVELOPMENT PROCESS

THE PANTHER was a direct response to the lurking presence of the Soviet T-34 and KV-1 tanks which were first encountered on 23 June 1941. The T-34 was relatively easy to produce and was soon available in large numbers. The KV-1, although rarer, easily outclassed the existing Panzer III and IV of the *Panzertruppen*. At the insistence of General Heinz Guderian, a special *Panzerkommision* was dispatched to the Eastern Front to assess the T-34. Among the features of the Soviet tank which the *Panzerkommision* considered most significant were the sloping frontal armour, which gave much improved shot deflection and also increased the effective armour thickness against penetration, the wide tracks of the T-34 also provided excellent mobility over soft ground.

This photograph showing shows the production of the wide tracks for the Panther, which as with the T-34, enabled the Panther to have improved mobility over soft ground.

The 76.2 mm main gun of the T-34 in contrast to the short barrelled howitzer type of the Mark IV had a reasonably high muzzle velocity making for good armour penetration, furthermore it also fired an effective high explosive round. This type of main gun was therefore considered by the *Panzerkommision* to be the new minimum standard for the next generation of German tanks. However, it should be noted that the Germans already boasted a comparable main gun which was fitted to Mark IV F2 tanks from 1942 onwards. Other than the massive frontal armour of the Panther the Mark IV F2, although approaching the limits of the design, actually incorporated many of the features which the Panther was intended to deliver.

In November 1941, the decision to up-gun the Panzer IV to the 50 mm gun had been dropped, and instead Krupp was contracted in a joint development to modify Rheinmetall's pending 75 mm anti-tank gun design, later known as the PaK 40 L/46. As the recoil length was considered too long for PaK 40 to be mounted in the Mark IV turret, the recoil mechanism and chamber had to be shortened in order that the weapon could serve as an effective *Kampfwagenkanone*. The conversion work was undertaken with the usual war time speed resulted in the new 75 mm KwK 40 L/43. This new tank gun, when firing an armour-piercing shot, achieved a dramatic rise in muzzle velocity was increased from 430 metres per second to 990 metres per second. Initially, the gun was mounted with a single-chamber, ball-shaped muzzle brake, which provided just under 50% of the recoil system's breaking ability. Firing the *Panzergranate* 39, the KwK 40 L/43 could penetrate 77 mm of steel armour at a range of 1,830 metres. This new *Kampfwagenkanone* was first introduced into the 1942 Panzer IV Ausf. F2 which was the equal of the T-34 in combat. A simple process of evolution and more efficient manufacturing programme could therefore have provided the hard pressed *Panzerwaffe* in 1942 with a large volume of tried and tested machines. In reality much needed resources were diverted to the Panther programme which did not bear fruit until 1944.

On 1st March 1943 Guderian, who had spent the previous year in the wilderness, was rehabilitated and appointed Inspector-General of the armoured Troops. His responsibilities were to determine armoured strategy and to oversee tank design and production and the training of Germany's panzer forces. According to Guderian, Hitler was far too easily persuaded to field a surfeit of new tank and tank destroyer designs, this resulted in unnecessary supply, logistical, and repair problems for the German forces in Russia. Guderian preferred the simple expedient of fielding large numbers of Panzer IIIs and Panzer IVs over smaller numbers of heavier tanks like the Panther which had limited range and required its own logistical channel of Panther spares. Guderian famously summed up these frustrations with the remark that *"logistics are the ball and chain of the tank forces"*.

The importance of an adequate supply of spare parts is often over looked when armoured affairs are considered, but without replacement parts tank formations soon grind to a halt. As Guderian was all too aware, the decision to deploy the Panther alongside the Mark III, IV and Tiger I added a real logistical headache. The introduction of the Panther meant that a system for the supply of an entire new set of spare parts had to be set up and drawn along a 2000 mile supply line as opposed to simply utilising the existing chain which provided Mark IV spares.

Despite the fact that, in Guderian's view, by utilising the existing Mark III and IV an acceptable battlefield solution lay within the grasp of the *Panzertruppen*, it was to prove fortunate that the Panther with the new longer barrelled L/70 was already under development. Had the tank not been a pet project enjoying Hitler's blessing, Guderian may well have prevailed in his desire to limit the number of German tank models. However, as a result of the inevitable evolution which was taking place on the battlefield, the presence of the Panther with its heavy hitting

Some graphic advice on how essential maintenance could avoid breakdowns from the Pantherfibel.

punch would soon be required and welcomed. The Russian T-34 equipped with an 85mm gun first appeared in late 1942. This up-gunned version of the T-34 was known to the Germans as the T-43 and it began to appear in large numbers during 1943.

The race towards producing tanks with a larger main gun was all about achieving increased muzzle velocity; the speed a projectile achieves at the moment it leaves the muzzle of the gun. Longer barrels give the propellant force more time to develop the speed of the shell before it leaves the barrel. For this reason longer barrels generally provided higher velocities and hence more penetrating power. During World War II the constant introduction of faster burning propellant, improved shells and longer barrel length led to constantly enhanced armour piercing capabilities on both sides.

Providing the antidote to the T-43 and T-43 required the introduction of a tank equipped with ideally the 8.8 cm KwK 36 L/56 of the Tiger I or alternatively the long barrelled 7.5 cm L/70.

There was no prospect of fitting either *Kampfwagenkanone* into the turret of the Mark IV. It was fortunate therefore that, driven on by Hitler's mania for radical new solutions, in late November 1941 Daimler-Benz (DB) and Maschinenfabrik Augsburg-Nürnberg AG (MAN) were each given the task of designing a completely new 30- to 35-ton tank. This development machine was designated VK30.02 and was required to was to be ready by April 1942 in time to be shown to Hitler for his birthday.

The Daimler Benz design imitated the T-34 almost completely. It closely resembled the T-34 in both hull and turret form and therefore posed obvious problems concerning battlefield recognition. Unusually for a German design it also incorporated a rear sprocket drive. The initial road wheel arrangement was also visually similar to that of the T-34 although Daimler's design initially used a leaf spring suspension whereas the T-34 incorporated coil springs. The Daimler Benz turret was smaller than that of the MAN design and had a smaller turret ring which was the result of the narrower hull required by the leaf spring suspension which lay outside of hull. The main advantages of the leaf springs over a torsion bar suspension were a lower hull silhouette and a simpler shock damping design. Unlike the T-34, the Daimler Benz design had the advantage of a three-man turret crew comprising commander, gunner, and loader whereas the T-34 turret initially allowed for only the commander and gunner. The planned L/70 75 mm gun was much longer and heavier than the L/43 and mounting it in the Daimler-Benz turret was difficult. Active consideration was given to reducing the turret crew to two men to address this problem but this retrograde step was eventually dropped.

The MAN design for the Panther embodied more conventional German thinking with the transmission and drive sprocket in the front and a turret placed centrally on the hull. One of the main design flaws of the Tiger lay in the low positioning of the front sprocket which made obstacle crossing less efficient. The design for the Panther eliminated this draw back by situating the front drive sprockets much higher than the road wheels making obstacle crossing more efficient and improving cross country performance in muddy conditions.

The MAN design for the Panther incorporated a Maybach petrol engine and eight torsion-bar suspension axles per side. Because of the torsion bar suspension and the drive shaft running under the turret basket, the MAN Panther needed to be higher and had a wider hull than the Daimler Benz design. The MAN Panther incorporated the "slack-track" Christie-style pattern of large road wheels with no return rollers for the upper run of track. Like the Tiger the main road wheels were interleaved but were arranged in just two rows eliminating the worst aspects of the Tiger I design.

The two designs were reviewed over a period from January through March 1942. Following a significant programme of development in which the running gear of the Daimler Benz design was altered to match that of the MAN design Reichminister Todt,

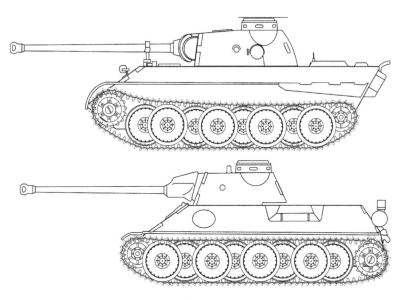

The VK30.02 prototype proposed by MAN (upper) seen alongside the alternative design proposed by Daimler-Benz (lower).

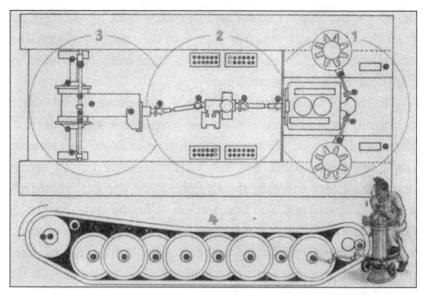

The complex drive arrangement of the Panther as shown in the Pantherfibel.

and later, his replacement Albert Speer, both recommended the Daimler Benz design to Hitler because of its several advantages over the initial MAN design.

However, by the final submission, MAN had substantially improved their design, having incorporated the best elements of Daimler Benz proposal. A final review by a special commission appointed by Hitler in May 1942 belatedly settled on the MAN design. Hitler approved this decision after reviewing it overnight. One of the principal reasons given for this decision was that the MAN design used an existing turret designed by Rheinmetall-Borsig, while the Daimler Benz design would have required the tooling a brand new turret to be designed and produced which would have significantly delayed the commencement of production.

Albert Speer recalled the trials and tribulations concerning the development of the Panther in his autobiography *Inside the Third Reich*. It produces a primary insight into Hitler's hands on involvement in the Panther design process.

"Since the Tiger had originally been designed to weigh fifty tons but as a result of Hitler`s demands had gone up to seventy five tons, we decided to develop a new thirty ton tank whose very name, Panther, was to signify greater agility. Though light in weight, its motor was to be the same as the Tiger`s, which meant it could develop superior speed. But in the course of a year Hitler once again insisted on clapping so much armour on it, as well as larger guns, that it ultimately reached forty eight tons, the original weight of the Tiger."

- CHAPTER 5 -
THE PANTHER IN PRODUCTION

THE MAN design had better fording ability, easier gun servicing and higher mobility due to better suspension, wider tracks, and a bigger fuel tank. A mild steel prototype of the Panther was produced by September 1942 and, after testing at Kummersdorf, was officially accepted. It was ordered to be placed into immediate production. The start of production was delayed, however, mainly because there were too few specialized machine tools needed for the machining of the hull. Finished tanks were produced in December and suffered from reliability problems as a result of this haste. The demand for this tank was so high that the manufacturing was soon expanded beyond MAN to include Daimler-Benz, Maschinenfabrik Niedersachsen-Hannover (MNH) and Henschel & Sohn in Kassel.

The initial production target was 250 tanks per month at MAN. This ambitious output target was increased to 600 per month in January 1943. Despite the maximum effort by the German war industry, including the use of slave labour, this figure was never reached. This was primarily as a result of disruption by Allied bombing, manufacturing bottlenecks, and other difficulties although it is possible that the target was simply too ambitious to be achieved even under favourable circumstances.

The use of slave labour in the complex production and logistical system which produced the Panther should not be overlooked. During the course of World War II the Germans abducted approximately 12 million people from almost twenty European countries; about two thirds of whom came from the Eastern Europe. These forced labourers provided the bulk of the labour in many of the German firms who supplied components and munitions for the Panther programme. Many of these

A Panther tank production line.

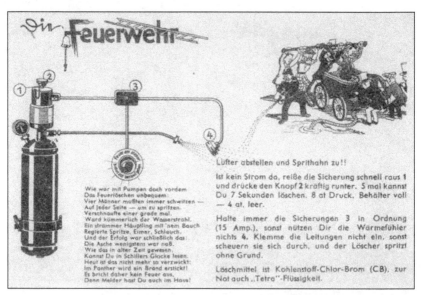

Fire was a constant threat to the Panther and in the Pantherfibel a strong emphasis was placed on fire fighting measures.

workers died as a result of their inhuman living conditions, mistreatment, malnutrition, or exhaustion and so became civilian casualties of war. At its peak the forced labourers comprised 20% of the German work force. Counting deaths and turnover, about 15 million men and women were forced labourers at one point or another during the war.

It was by resorting to such measures that Panther production was maintained at any level but the projected figure of 600 per month remained a pipe dream. In 1943 production averaged 148 per month. In 1944, it averaged out at 315 a month with 3,777 machines being built in that year. Panther production peaked at 380 in July 1944 and ended around the end of March 1945, by which time at least 6,000 built in total. Front-line combat strength peaked on 1st September 1944 at 2,304 tanks, but that same month a record number of 692 Panther tanks were reported lost.

Both the Tiger and the Panther used Maybach engines so it was no surprise that Allied air forces were soon targeting the Maybach engine plant. This plant was first bombed the night of 27/28th April 1944 and the attack was so severe and accurate that production was completely shut down for five months. Fortunately for the *Panzerwaffe* this exigency had already been anticipated and a second plant had already been planned, the Auto-Union plant at Siegmar, and this came online in May 1944 ensuring continuity of supply.

Targeting of the factories which produced the Panther itself began with a bombing raid on the Daimler Benz plant on 6th August 1944, a follow up raid took place on the night of 23/24th August 1944. MAN was first struck on 10th September then again in rapid succession on 3rd October and 19th October 1944. Despite these attacks however, production was soon resumed at MAN and the allied air forces returned to the fray on 3rd January 1945 and finally on 20/21st February 1945. The MNH was not attacked until very late in the war and was not targeted until 14th March. There was one final raid on 28th March 1945.

In addition to interfering with tank production goals, the bombing forced a steep drop in the production of spare parts.

Unlike the risqué illustrations in the Tigerfibel the young ladies in the Pantherfibel were all depicted with clothes on.

Spare parts as a percentage of tank production dropped from 25–30 per cent in 1943, to 8 per cent in the fall of 1944. This only compounded the problems with reliability and numbers of operational Panthers, as tanks in the field had to be cannibalized for parts.

PRODUCTION FIGURES

The Panther was the third most numerous German armoured fighting vehicle.

Production by type			
Model	**Number**	**Date**	**Notes**
Prototype	2	11/42	Designated V1 and V2
Ausf. D	842	1/43 to 9/43	
Ausf. A	2,192	8/43 to 6/44	Sometimes called Ausf. A2
Ausf. G	2,953	3/44 to 4/45	
Befehlspanzer Panther	329	5/43 to 2/45	Converted
Beobachtungspanzer Panther	41	44 to 45	Converted
Bergepanther	347	43 to 45	

Panther production in 1944 by manufacturer	
Manufacturer	**% of total**
Maschinenfabrik Augsburg-Nürnberg (M.A.N.)	35%
Daimler-Benz	31%
Maschinenfabrik Niedersachsen-Hannover	31%
Other	3%

- CHAPTER 6 -
THE PANTHER IN COMBAT

PANTHERS WERE first supplied to form *Panzer Abteilung 51* (tank Battalion 51) on 9 January, and then Pz.Abt. 52 on 6 February 1943.

The first production Panther tanks were plagued with mechanical problems. The engine was dangerously prone to overheating and suffered from connecting rod or bearing failures. Gasoline leaks from the fuel pump or carburettor, as well as motor oil leaks from gaskets easily produced fires in the engine compartment; several Panthers were destroyed in such fires. Transmission and final drive breakdowns were the most common and difficult to repair. A large list of other problems were detected in these early Panthers, and so from April through May 1943 all Panthers were shipped to Falkensee and Nuremburg for a major rebuilding program. This did not correct all of the problems, so a second program was started at Grafenwoehr and Erlangen in June 1943.

Panther tanks move into formation during Operation Citadel. These machines were part of the initial run of 200 tanks deployed in July 1943.

PANTHER ON THE EASTERN FRONT

The Panther tank was viewed by Hitler as an essential component of the forthcoming Operation *Zitadelle*, and on his orders the attack was delayed several times because of the mechanical problems which were still being encountered. The eventual start date of the battle was delayed as long as possible and commenced only six days after the last of the 200 Panthers had been delivered to the front. This hurriedness resulted in major problems in Panther units during the Battle of Kursk, as in addition to all of the other problems, tactical training at the unit level, co-ordination by radio, and driver training were all seriously deficient.

It was not until 29th June, 1943, that the last of a total of 200 rebuilt Panthers were finally issued to Panther Regiment von Lauchert, of the XLVIII Panzer Corps (4 Panzer Army). Two were immediately lost due to motor fires upon disembarking from the trains. By 5th July, when the Battle of Kursk started, there were only 184 operational Panthers. Within two days, this total had dropped to just 40. On 17th July 1943 after Hitler had ordered a stop to the German offensive, Gen. Heinz Guderian sent in the following preliminary assessment of the Panthers:

"Due to enemy action and mechanical breakdowns, the combat strength sank rapidly during the first few days. By the evening of 10 July there were only 10 operational Panthers in the front line. 25 Panthers had been lost as total write offs (23 were hit and burnt and two had caught fire during the approach march). 100 Panthers were in need of repair (56 were damaged by hits and mines and 44 by mechanical breakdown). 60 percent of the mechanical breakdowns could be easily repaired. Approximately 40 Panthers had already been repaired and were on the way to the front. About 25 still had not been recovered by the repair service. On the evening of 1th July, 38 Panthers were operational, 31 were total write offs and 131 were in need of repair. A slow increase in the combat strength is observable.

A Panther is left destroyed after the Battle of Kursk.

The large number of losses by hits (81 Panthers up to 10th July) attests to the heavy fighting."

During *Zitadelle* the Panthers claimed a total 267 destroyed Soviet tanks. Given the circumstances this was a remarkable achievement and pointed towards the fact that under the right circumstances this was could have been a very impressive design indeed.

A later report on 20th July 20, 1943 showed 41 Panthers as operational, 85 as repairable, 16 severely damaged and needing repair in Germany, 56 burnt out (due to enemy action), and 2 that had been completely destroyed by motor fires at the railhead.

Even before the Germans ended their offensive at Kursk, the Soviets began their counteroffensive and succeeded in pushing the Germans back into a steady retreat. As a result of the headlong withdrawal many Panthers could not be recovered and had to be left on the battlefield This was to lead to a drastic decline in operational numbers and a report 11th August 1943 showed that the numbers of total write offs in the Panther force swelled to 156, with only 9 operational. The German Army was forced into a fighting retreat and increasingly lost Panthers in combat as well as from abandoning and destroying damaged vehicles.

The Panther demonstrated its capacity to destroy any Soviet tank from long distance during the Battle of Kursk, and had a very high overall kill ratio. However, it comprised less than seven per cent of the estimated 2,400–2,700 total tanks deployed by the Germans in this battle, and its effectiveness was limited by its mechanical problems and the in-depth layered defence system of the Soviets at Kursk. Ironically the greatest contribution to this titanic battle may have been a highly negative one. Hitler's decisions to delay the original start of Operation *Zitadelle* for a total of two months was at least partially due to his desire to see the panther in action. The precious extra time was used by the Soviets to build up an enormous concentration of minefields, anti-tank guns, trenches and artillery defences which ultimately thwarted the German ambitions.

After the losses of the Battle of Kursk, the German Army was forced into a constant state of retreat before the advancing Red Army. The numbers of Panthers were slowly re-built on the Eastern Front, and the operational percentage increased as its reliability was improved. In March 1944, Guderian reported:

A Panther photographed on the Eastern Front in 1944.

Panther tanks of the Großdeutschland Division advance in the area of Iaşi, Romania in 1944.

"Almost all the bugs have been worked out". Despite his bold words many units continued to report significant mechanical problems, especially with the final drive. There undoubtedly were some real advances in reliability and the greatly outnumbered Panthers for the remainder of the war were used as mobile reserves to fight off major attacks.

THE CONTEMPORARY VIEW NUMBER 2
"NEW HEAVY TANK: THE Pz. Kw. 5 (PANTHER)"
FROM INTELLIGENCE BULLETIN, JANUARY 1944
NEW HEAVY TANK: THE Pz. Kw. 5 (PANTHER)

When the Pz. Kw. 6 (Tiger) became standard, the Pz. Kw. 5 (Panther) was still in an experimental stage. Now that the Panther has joined the German tank series as a standard model, a general description of this newest "land battleship" can be made available to U.S. military personnel. Much of the data presented here comes from Russian sources, inasmuch as the Pz. Kw. 5 has thus far been used only on the Eastern Front.

The Panther is a fast, heavy, well-armoured vehicle. It mounts a long 75-mm gun. Weighing 45 tons, the new tank appears to be of a type intermediate between the 22-ton Pz. Kw. 4 and the 56-ton Pz. Kw. 6. The Panther has a speed of about 31 miles per hour. It corresponds roughly to our General Sherman, which the Germans have always greatly admired. It is believed that the 75-mm gun is the Kw.K.. This tank gun is a straight-bore weapon with a muzzle brake, and has an over-all length of 18 feet 2 inches.

Although equipped with the same motor as the Tiger, the Panther has lighter armour and armament. For this reason it is capable of higher speed and greater maneuverability. The Panther is also provided with additional armour plate, 4- to 6-mm thick, (not shown in fig. 1) along the side, just above the suspension wheels and the sloping side armour plate.

When a flexible tube with a float is attached to the air intake, the Panther has no difficulty in fording fairly deep streams. There is a special fitting in the top of the tank for attaching this tube.

Like the Pz. Kw. 6's, the Pz. Kw. 5's are organized into separate tank battalions. During the summer of 1943, the Germans used many of these new tanks on the Russian front.

Although the Russians have found the Pz. Kw. 5 more manoeuvrable than the Pz. Kw. 6, they are convinced that the new tank is more easily knocked out. Fire from all types of rifles and machine guns directed against the peep holes, periscopes, and the base of the turret and gun shield will blind or jam the parts, the Russians say. High explosives and armour-piercing shells of 54-mm (2.12 inches) calibre, or higher are effective against the turret at ranges of 875 yards or less. Large-calibre artillery and self-propelled cannon can put the Panther out of action at ordinary distances for effective fire. The vertical and sloping plates can be penetrated by armour-piercing shells of 45-mm (1.78 inches) calibre, or higher. Incendiary armour-piercing shells are said to be especially effective, not only against the gasoline tanks, but against the ammunition, which is located just to the rear of the driver.

The additional armour plate above the suspension wheels is provided to reduce the penetration of hollow-charge shells. According to the Russians, it is ineffective; antitank grenades, antitank mines, and Molotov cocktails are reported to be effective against the weak top and bottom plates and the cooling and ventilating openings on top of the tank, just above the motor.

However, it should definitely be stated that the Pz. Kw. 5 is a formidable weapon—a distinct asset of the German Army.

1. With certain alterations the Pz. Kw. 6 may weigh as much as 62 tons. For an illustrated discussion of the Pz. Kw. 6, see Intelligence Bulletin, Vol. I, No. 10, pp. 19-23.

2. Kampfwagenkanone-tank gun.

The missile defeating qualities of sloped armour is graphically demonstrated in this page from the Pantherfibel.

The Panzer Mark V seen here in combat accompanied by infantry support.

The highest total number of operational Panthers on the Eastern Front was achieved in September 1944, when some 522 were listed as operational out of a total of 728 machines. Throughout the rest of the war, Germany continued to deploy the majority of available Panther forces on the Eastern Front. The last recorded status, on 15th March 1945, listed 740 Panthers on the Eastern Front with 361 operational. By this time the end was in sight as the Red Army had already entered East Prussia and was advancing through Poland.

In August 1944 Panthers were deployed in Warsaw during the uprising as a mobile artillery and troops support. At least two of them were captured in the early days of the conflict and used in actions against Germans, including the liberation of Gęsiówka concentration camp on 5th August, when the soldiers of *"Wacek"* platoon used the captured Panther (named *"Magda"*) to destroy the bunkers and watchtowers of the camp. Most of the Germans in the camp were killed; insurgents had lost two people and liberated almost 350 people. After several days they were immobilized due to the lack of fuel and batteries and were set ablaze to prevent them from being re-captured by the German forces.

The organisation of Panther battalions varied but an optimal organisation is set out below. In practice the number of operational machines was never achieved.

- Battalion Command
  (Composed of Communication and Reconnaissance platoons)
- Communication Platoon -
  3 × *Befehlswagen* Panther SdKfz.267/268
- Reconnaissance Platoon - 5 × Panther
- 1st Company - 22 × Panther
  - Company Command - 2 × Panther
    - 1st Platoon - 5 × Panther
    - 2nd Platoon - 5 × Panther
    - 3rd Platoon - 5 × Panther
    - 4th Platoon - 5 × Panther
- 2nd Company - 22 × Panther (composed as 1st Company)
- 3rd Company - 22 × Panther (composed as 1st Company)
- 4th Company - 22 × Panther (composed as 1st Company)
- Service Platoon - 2 × *Bergepanther* SdKfz.179

From 3rd August 1944, the new Panzer-Division 44 organisation called for a Panzer division to consist of one Panzer regiment with two Panzer battalions – one of 96 Panzer IVs and one of 96 Panthers. Actual strengths tended to differ, and in reality were far lower after combat losses were taken into account.

- CHAPTER 7 -
PANTHERS IN COMBAT ON THE WESTERN FRONT - FRANCE

A T THE time of the invasion of Normandy, there were initially only two Panther-equipped Panzer regiments on the entire Western Front, they fielded a total of 156 Panthers. From June through August 1944, an additional seven Panther regiments were sent into France, reaching a maximum strength of 432 in a status report dated 30th July , 1944.

The majority of German Panzer forces in Normandy – six and a half divisions, were stationed around the vital town of Caen facing the Anglo-Canadian forces of the 21st Army Group; and the numerous battles to secure the town became collectively known as the Battle of Caen. While there were sectors of heavy bocage around Caen, there were also many open fields over which the Allied armour had to attack. This allowed the Panther to play to its strengths and engage the attacking enemy armour at long range. By the time of the Normandy Campaign however, British

Panthers in a French village, Summer 1944.

Panther in bocage, Summer 1944.

Divisional Anti-tank Regiments were well equipped with the excellent 17 pounder gun (the 17 pounder also replaced the US gun on some M10 tank Destroyers in British service), making it equally as perilous for the Panthers to launch attacks across these same killing fields. The British had begun converting regular M4 Shermans to carry the 17 pounder gun (nicknamed Firefly) prior to the D-day landings, and while limited numbers meant that during Normandy not more than one Sherman in four was of the Firefly variant, the lethality of its gun against German armour made them priority targets for German gunners.

US forces in the meantime, facing one and a half German Panzer divisions, mainly the Panzer Lehr Division, struggled in the heavy, low-lying bocage terrain west of Caen. Against the M4 Shermans of the Allied tank forces during this time, the Panther tank again proved to be most effective when fighting in open country and firing at long range - its combination of superior armour and firepower allowed it to engage at distances from which the Shermans could not respond. However, the Panther struggled in the enclosed bocage country of Normandy, and was vulnerable to side and close-in attacks in the built-up areas of cities and small towns. The commander of the Panzer *Lehr* Division, Gen. Fritz Bayerlein, reported the weaknesses of the Panther tank in the fighting in Normandy in a very damning report:

"While the PzKpfw IV could still be used to advantage, the PzKpfw V [Panther] proved ill adapted to the terrain. The Sherman because of its manoeuvrability and height was good... [the Panther was] poorly suited for hedgerow terrain because of its width. Long gun barrel and width of tank reduce manoeuvrability in village and forest fighting. It is very front-heavy and therefore quickly wears out the front final drives, made of low-grade steel. High silhouette. Very sensitive power-train requiring well-trained drivers. Weak side armour; tank top vulnerable to fighter-bombers. Fuel lines of porous material that allow gasoline fumes to escape into the tank interior causing a grave fire hazard. Absence of vision slits makes defence against close attack impossible."

The subject of how to avoid anti-tank defences was an important consideration and was represented graphically in the Pantherfibel.

A pair of Panthers rendered useless after they have been knocked-out and left at the roadside, Normandy, Summer 1944.

Through September and October, a series of new Panzerbrigades equipped with Panther tanks were sent into France to try to stop the Allied advance with counterattacks. This culminated in the Battle of Arracourt (September 18–29, 1944), in which the mostly Panther equipped German forces suffered heavy losses fighting against the 4th armoured Division of Patton's 3rd Army, which were still primarily equipped with 75 mm M4 Sherman tanks and yet came away from the battle with only a few losses. The Panther units were newly formed, poorly trained, and tactically disorganized; most units ended up stumbling into ambush situations against seasoned U.S. tank crews.

- CHAPTER 8 -

WESTERN FRONT -
ARDENNES OFFENSIVE

A STATUS report on 15th December, 1944 listed a record high of 471 Panthers deployed the Western Front, with 336 operational accounting for a healthy 71 per cent of the available force. This was one day before the start of the Battle of the Bulge; 400 of the tanks assigned to the Western Front were in units detailed for the offensive.

During the Battle Of The Bulge The Panther once again demonstrated its prowess in open country, where it could destroy its victims at long range with near-impunity. The reverse side of the coin was once again in evidence as the vulnerability of the Panther in the close-in fighting of the small towns of the Ardennes, was cruelly exposed and there were consequently very heavy losses. A status report on January 15, 1945 showed only 97 operational Panthers left in the units involved in the operation, out of 282 still in their possession. Total write-offs were listed as 198.

A burnt out Panther Ausf. G at the Battle of the Bulge, which has been penetrated in the sponson.

The Operation *Greif* commando mission included five Panthers assigned to Panzerbrigade 150, disguised to look like M10 tank Destroyers by welding on additional plates, applying US-style camouflage paint and markings. This was carried out as part of a larger operation that involved soldiers disguised as Americans and other activities. The disguised Panthers were detected and destroyed and their story was reported in the US intelligence magazine *Tactical and Technical Trends* No. 57, April 1945.

THE CONTEMPORARY VIEW NUMBER 3

GERMANS DISGUISE PANTHERS

CLEVERLY IMITATE M10 GUN CARRIAGE

Investigation of four German Panther tanks knocked out in the Malmedy area in the December breakthrough in Belgium revealed that the tanks were carefully and cleverly disguised as U.S. M10 gun motor carriages.

Panther disguised as an M10 Tank Destroyer.

After inspecting the tanks and realizing the amount of time, work, and materials involved in order to imitate the appearance of the M10, Ordnance intelligence investigators expressed the opinion that these disguised tanks, used in the proper tactical situation and at the proper time, would have caused considerable damage.

Because the false vehicle numbers of the tanks knocked out were B-4, B-5, B-7, and B-10, investigators concluded that at least ten similarly disguised tanks might have been in action.

Inside the one tank which was not blown up too badly to be inspected were found items of U.S. clothing such as a helmet, overcoat, and leggings. To heighten the deception, U.S. stars were painted on both sides and also on the top of the turret, the entire tank was painted O.D., and U.S. unit markings were painted on the false bow and rear.

In disguising the Panther the distinctive cupola was removed from the turret and two semicircular hatch covers were hinged in its place to the turret top in order to cover the opening. In addition, it was necessary to remove extra water cans, gas cans, the rammer staff container, and other external accessories.

The tank then was camouflaged or disguised with sheet metal, that used on the turret and upper bow being three twenty-seconds of an inch thick and that on time sides of the hull being nine sixty-fourths of an inch thick. The lower part of the false bow was thicker, possibly made of double plates. To accomplish the deceptive modifications, which pointed to at least fourth or fifth echelon alterations, the work probably was done by maintenance units rather than at a factory. The work probably was divided into four sections: turret, bow, rear, and sides.

Top view of Panther tank disguised as U.S. M10 gun carriage, showing hatch covers used in place of cupola.

TURRET CHANGES

The turret was disguised by using five pieces of sheet metal, two of which were cut to resemble the distinctive sides of the M10 turret and then were flanged on the edges, bent to shape, and stiffened with small angle iron. The gun shield was carefully formed from another sheet to the exact shape of the M10 shield, and a hole was made to the right of the gun hole in the shield for the co-axial M.G. 34, a hole which does not exist in the M10 shield. Two pieces of sheet metal made up the rear of the turret, one representing the bottom slant surface of the rear and one representing the counterweight. The pieces representing the sides and rear were joined together and braced with angle iron, and the whole was attached to the turret. The false gun shield was attached to the Panther gun shield, and all the lifting rings, brackets, extra-armour studs, etc., found on the M10 turret were carefully duplicated and welded to the false turret.

Left front view with turret reversed. Note false final-drive housing at bottom of bow and false side apron.

FALSE BOW

Approximately four pieces of sheet metal, shaped to imitate as closely as possible the contours of the M10 bow, made up the false bow, necessary because the Panther bow is bulkier than the M10. The false bottom was shaped to give the characteristic appearance of the front drive sprocket housing of the M10, and the top was shaped carefully and various component pieces attached to the front of the tank. All the brackets, lifting rings, towing devises, etc., of the M10 bow were also imitated. A square opening was cut in the false bow to permit the use of the bow M.G. 34, but a removable cover attached with a small chain was made for this opening.

FALSE REAR AND SIDES

The false rear was made of sheet metal. It was a faithful duplicate of the M10 rear except for two holes to permit the twin exhaust elbows of the Panther to protrude.

An attempt was made to imitate the skirting armour of the M10 which appears to hang lower than the side armour of the Panther and is bevelled in at the bottom. A long flat strip of sheet metal was attached to the sides parallel to the ground, and a vertical sheet strip was attached at right angles to this strip to give the appearance of low skirting armour.

Front view showing plate over machine-gun opening, false lifting rings and brackets, and markings.

Features which aid in recognizing disguised Panthers and which cannot be camouflaged easily are:

1. The distinctive Panther bogie suspension. (The M18 motor gun carriage now has a somewhat similar suspension.)
2. The muzzle brake on the 7.5 cm Kw.K. 42.
3. The wide and distinctive track of the Panther tank.

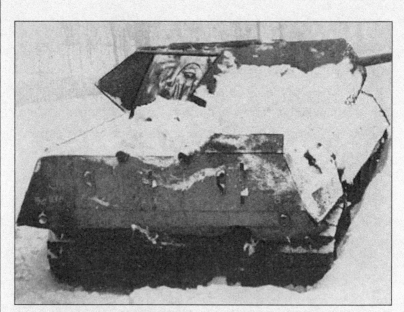

Rear view showing false tail plate. Note exhausts and dummy fittings.

In February 1945, eight Panzer divisions with a total of 271 Panthers were transferred from the West to the Eastern Front. Only five Panther battalions remained in the west.

One of the top German Panther commanders was SS-*Oberscharführer* Ernst Barkmann of the 2nd SS-Panzer Regiment "*Das Reich*". By the end of the war, he had some 80 tank kills claimed.

CHAPTER 9

BUILDING THE PANTHER

THE COST of a Panther tank was 117,100 Reichmarks (RM). This compared favourably with 82,500 RM for the StuG III, 96,163 RM for the Panzer III, 103,462 RM for the Panzer IV, and 250,800 RM for the Tiger I. These figures did not include the cost of the armament and radio. Expressed in terms of Reichmarks per ton the Panther tank was arguably one of the most cost-effective German tanks of World War II. However, these cost figures should be understood in the context of the time period in which the various tanks were first designed, as the Germans armaments industry increasingly strove for designs and production methods that would allow for higher production rates, and thereby steadily reduced the cost of their tanks.

The process of streamlining the production of German tanks first began after Speer became Reichminister in early 1942, and steadily accelerated through 1943 reaching a peak in 1944;

Panther with track segments hung on the turret sides to augment the armour, 1944.

production of the Panther tank therefore coincided with this period of increased manufacturing efficiency.

In the pre-war era German tank manufacturers relied heavily on a large pool of skilled and willing workers. Even after the outbreak of World War II the armaments industry continued to utilize heavily labour-intensive and costly manufacturing methods unsuited to mass production. Under the influence of Albert Speer the increasing use of forced labour and increased production efficiencies led to a jump in output; although it should be noted that, even with streamlined production methods and slave labour, Germany could not hope to approach the efficiency of Allied manufacturing during World War II.

Initial production Panthers had a face-hardened frontal armour which formed the glacis plate the benefits of face-hardening was that it caused uncapped rounds to shatter, but as capped armour-piercing capped rounds became the standard in all armies this expensive and difficult process was no longer relevant and the requirement was deleted on 30th March 1943. By August 1943, Panthers were being built only with a homogeneous steel glacis plate which helped to bring down costs and speed up production.

Although the front hull of the panther boasted only 80 mm of armour as opposed to the 100 mm of the Tiger I, the fact that the armour sloped back at 55 degrees from the vertical, gave it additional advantages and effectively produced the same benefits as the thicker Tiger armour. In addition the front glacis plates were welded and interlocked for additional strength. The combination of a steep slope and thick armour meant that few Allied or Soviet weapons could hope to penetrate the Panther frontally other than at very close ranges.

It was an altogether different matter with regard to the side armour. In order to achieve the weight savings which allowed the Panther to function at all the armour for the side hull and superstructure however was much thinner at just 40–50 mm. The thinner side armour was essential to keep the overall weight within reasonable bounds, but it made the Panther extremely vulnerable to attacks from the side at relatively long ranges by most Allied and Soviet tank and anti-tank guns. German tactical doctrine for the use of the Panther thus emphasized the importance of flank protection. Five millimetre thick skirt armour, known as *Schürzen*, was fitted to the sides of the Panthers. This flimsy addition was intended to provide protection for the lower side hull from Soviet anti-tank rifle fire and was fixed on the hull side by means of a series of brackets. In the rough conditions encountered in the field these plates were constantly being torn off and many surviving pictures show Panthers missing these side panels.

Zimmerit coating against Soviet magnetic mines was applied at the factory on late Ausf. D models commencing in September 1943; an order for field units to apply Zimmerit to older versions of the Panther was issued in November 1943. However in September 1944, these orders were countermanded and a new to stop all application of Zimmerit were issued. This new order was based on combat reports that hits on the Zimmerit had caused vehicle fires.

Panther crews were aware of the weak side armour and made unauthorized augmentations by hanging track links or spare road wheels onto the turret and the hull sides. The rear hull top armour was soon recognised as the extreme weak point of the Panther it was only 16 mm thick, and housed two radiator fans and four air intake louvres over the engine compartment. This made the Panther highly vulnerable to strafing attacks by aircraft. With such thin armour even those aircraft armed with just machine guns were potentially dangerous opponents. The Panther was also highly vulnerable to shrapnel damage from airbursts.

As the war progressed, Germany was forced to curtail the use of certain critical alloy materials in the production of armour plate, such as nickel, tungsten, molybdenum, and manganese. The loss of these alloys resulted in substantially reduced impact resistance levels compared to earlier armour. Manganese from mines in the Ukraine ceased when the German Army lost control of this territory in February 1944. Allied bombers struck the Knabe mine in Norway and stopped a key source of molybdenum; other supplies from Finland and Japan were also cut off. The loss of molybdenum, and its replacement with other substitutes to maintain hardness, as well as a general loss of quality control resulted in an increased brittleness in German armour plate, which developed a tendency to fracture when struck with a shell. Testing by U.S. Army officers in August 1944 in Isigny, France showed catastrophic cracking of the armour plate on two out of three Panthers examined.

PANTHER TURRETS AS FORTIFICATIONS

From 1943, Panther turrets were mounted in fixed fortifications, some were normal production models, but most were made specifically for the task, with additional roof armour to withstand artillery. Two types of turret emplacements were used;

Pantherturm fortification under inspection in Italy in June 1944.

70

(*Pantherturm* III - *Betonsockel* - concrete base) and (*Pantherturm* I - *Stahluntersatz* - steel sub-base). They housed ammunition storage and fighting compartment along with crew quarters. A total of 182 of these were installed in the fortifications of the Atlantic Wall and West Wall, 48 in the Gothic Line and Hitler Line, 36 on the Eastern Front, and 2 for training and experimentation, for a total of 268 installations by March 1945. They proved to be costly to attack, and difficult to destroy.

PANTHER BATALLION ORGANIZATION

From September 1943, one Panzer battalion with 96 Panthers comprised the Panzer regiment of a Panzer-Division 43.
- Battalion Command
  (Composed of Communication and Reconnaissance platoons)
- Communication Platoon -
  3 × *Befehlswagen* Panther SdKfz.267/268
- Reconnaissance Platoon - 5 × Panther
- 1st Company - 22 × Panther
    - Company Command - 2 × Panther
        - 1st Platoon - 5 × Panther
        - 2nd Platoon - 5 × Panther
        - 3rd Platoon - 5 × Panther
        - 4th Platoon - 5 × Panther
- 2nd Company - 22 × Panther (composed as 1st Company)
- 3rd Company - 22 × Panther (composed as 1st Company)
- 4th Company - 22 × Panther (composed as 1st Company)
- Service Platoon - 2 × *Bergepanther* SdKfz.179

From 3rd August 1944, the new Panzer-Division 44 organisation called for a Panzer division to consist of one Panzer regiment with two Panzer battalions – one of 96 Panzer IVs and one of 96 Panthers. Actual strengths tended to differ, and in reality were far lower after combat losses were taken into account.

THE SOVIET RESPONSE

The importance of the tank on the Eastern Front led to an arms race between the Germans and Soviets to produce tanks with ever greater armour and firepower. The Tiger I and Panther tanks were German responses to encountering the T-34 in 1941. Soviet firing tests against a captured Tiger in April 1943 showed that the T-34's 76 mm gun could not penetrate the front of the Tiger I at all, and the side only at very close range. An existing Soviet 85 mm antiaircraft gun, the 52-K, was found to be very effective against the frontal armour of the Tiger I, and so a derivative of the 52-K 85 mm gun (F-34 tank gun) was developed for the T-34. The Soviets thus had already embarked on the 85 mm gun upgrade path before encountering the Panther tank at the Battle of Kursk.

After much development work, the first T-34-85 tanks entered combat in March 1944. When tested by Wehrmacht, the production version of the T-34's new 85 mm F-34 gun proved to be ineffective against the Panther's frontal armour at the standard *Panzerwaffe* engagement range of 2,000 m, meaning the Soviet tanks were out-ranged in open country, while the Panther's main gun could penetrate the T-34 frontal armour at this range from any angle. Although the T-34-85 tank was not quite the equal of the Panther, it was much better than the 76.2 mm-armed versions and made up for its quality shortcomings by being produced in greater quantities than the Panther. New self-propelled anti-tank vehicles based on the T-34 hull, such as the SU-85 and SU-100, were also developed. A German Army study dated October 5, 1944 showed that from a 30 degree side angle the Panther's gun could easily penetrate the turret of the T-34-85 from the front at ranges up to 2000 m, and the frontal hull armour at 300 m, whereas from the front, the T-34-85 could only penetrate the non-mantlet part of the Panther turret by closing to a range of 500 m. From the side, the two were nearly equivalent as both tanks could penetrate the other from long range. T-34-85 production was soon varied to allow for the introduction of two replacement guns, the D-5T and ZiS-S-53, the later becoming a production standard for the rest of the war.

The Battle of Kursk convinced the Soviets of the need for even greater firepower. A Soviet analysis of the battle in August 1943 showed that a Corps artillery piece, the A-19 122 mm gun, had performed well against the German tanks in that battle, and so development work on the 122 mm equipped IS-2 began in late 1943. Soviet tests of the IS-2 versus the Panther included a claim of one shot that could penetrate the Panther from the front armour through the back armour. However, German testing showed that the 122 mm gun could not penetrate the glacis plate of the Panther at all, but it could penetrate the front turret/mantlet of the Panther at ranges up to 1500 m. At a 30 degree side angle the Panther's 75 mm gun could penetrate the front of the IS-2s turret at 800 m and the hull nose at 1000 m. From the side, the Panther was more vulnerable than the IS-2. Thus the two tanks, while nearly identical in weight, had quite different combat strengths and

Panzerbefehlswagen Panther Ausf. A (Sd.Kfz. 267) of the Panzergrenadier-Division Großdeutschland photographed in southern Ukraine in 1944.

A Panther tank is passing anti-tank obstacles of the Westwall near Weissenburg / Bergzabern, January 1945.

weaknesses. The Panther carried much more ammunition and had a faster firing cycle than the IS-2, which was a lower and more compact design; the IS-2s A-19 122 mm gun used a two piece ammunition which slowed its firing cycle.

THE AMERICAN AND BRITISH RESPONSE

The Western Allies' response was inconsistent between the Americans and the British. Although the western Allies were aware of the Panther and had access to technical details through the Soviets, the Panther was not employed against the western Allies until early 1944 at Anzio in Italy, where Panthers were employed in small numbers. Until shortly before D-Day, the Panther was thought to be another heavy tank that would not be built in large numbers. However, just before D-Day, Allied intelligence investigated Panther production, and using a statistical analysis of the road wheels on two captured tanks, estimated that Panther production for February 1944 was 270, thus indicating that it would be found in much larger numbers than had previously been anticipated. In the planning for the Battle of Normandy, the US Army expected to face a handful of German heavy tanks alongside large numbers of Panzer IVs, and thus had little time to prepare to face the Panther. Instead, 38% of the German tanks in Normandy were Panthers, whose frontal armour could not be penetrated by the 75 mm guns of the US M4 Sherman.

The British were more astute in their recognition of the increasing armour strength of German tanks, and by the time of the Normandy invasion their program that mounted the excellent 17-pounder anti-tank gun on some of their M4 Shermans had provided more than 300 of these Sherman Fireflies. The British lobbied during the war to use American production lines for building many Fireflies but these demands were ignored due to suspicion of British tank designs after they had done poorly in North Africa. There were also 200 interim Challenger tanks with the 17 pounder and improved tank designs under development. British and Commonwealth tank units in Normandy were initially

M4 Shermans in combat.

equipped at the rate of one Firefly in a troop with three Shermans or Cromwells. This ratio increased until, by the end of the war, half of the British Shermans were Fireflies. The Comet with a similar gun to the 17-pounder had also replaced the 75 mm gun Sherman in some British units. The 17-pounder with APCBC shot was more or less equivalent in performance to the Panther's 75 mm gun, but superior with APDS shot.

The US armour doctrine at the time was dominated by the head of Army Ground Forces, Gen. Lesley McNair, an artilleryman by trade, who believed that tanks should concentrate on infantry support and exploitation roles, and avoid enemy tanks, leaving them to be dealt with by the tank destroyer force, which were a mix of towed anti-tank guns and lightly armoured tanks with open top turrets with 3-inch (M-10 tank destroyer), 76 mm (M18 Hellcat) or later, 90 mm (M36 tank destroyer) guns. This doctrine led to a lack of urgency in the US Army to upgrade the armour and firepower of the M4 Sherman tank, which had previously performed well against the most common German tanks, the Panzer III and Panzer IV, encountered in Africa and Italy. As with the Soviets, the German adoption of thicker armour and the 7.5 cm KwK 40 in their standard tanks prompted the U.S. Army to develop the more powerful 76 mm version of the M4 Sherman tank in April 1944. Development of a heavier tank, the M26 Pershing, was delayed mainly by McNair's insistence on "battle need" and emphasis on producing only reliable, well-tested weapons, a reflection of America's 3,000 mile supply line to Europe.

An AGF (Armored Ground Forces) policy statement of November 1943 concluded the following:

"The recommendation of a limited proportion of tanks carrying a 90mm gun is not concurred in for the following reasons: The M4 tank has been hailed widely as the best tank of the battlefield today... There appears to be no fear on the part of our forces of the German Mark VI (Tiger) tank. There can be no basis for the

T26 tank other than the conception of a tank-vs-tank duel-which is believed to be unsound and unnecessary. Both British and American battle experience has demonstrated that the antitank gun in suitable numbers is the master of the tank... There has been no indication that the 76mm antitank gun is inadequate against German Mark VI tank."

U.S. awareness of the inadequacies of their M4 tanks grew only slowly. All U.S. M4 Shermans that landed in Normandy in June 1944 had the 75 mm gun. The 75 mm M4 gun could not penetrate the Panther from the front at all, although it could penetrate various parts of the Panther from the side at ranges from 400 to 2,600 m (440 to 2,800 yd). The 76 mm gun could also not penetrate the front hull armour of the Panther, but could penetrate the Panther turret mantlet at very close range. In August 1944, the HVAP (high velocity armour-piercing) 76 mm round was introduced to improve the performance of the 76 mm M4 Shermans. With a tungsten core, this round could still not

British Firefly in Namur, 1944.

penetrate the Panther glacis plate, but could punch through the Panther mantlet at 730 to 910 m, instead of the usual 90 meters for the normal 76 mm round. However, tungsten production shortages meant that this round was always in short supply, with only a few rounds available per tank, and some M4 Sherman units were not issued with any ammunition of this type.

Sherman tank shells used a high flash powder, making it easier for German crews to spot their opponents. German tanks conversely used a low flash powder making it harder for Allied crews to spot them. Due to the narrowness of their tracks which did little to spread the weight the Sherman also possessed an inferior cross country mobility in relation to the Panthers. This proved to be the case on all adverse surfaces from mud through to sheet ice. Meanwhile it is important to note that the Panther is around 15 tons heavier than the M4. Brig. Gen. J.H. Collier noted:

"I saw where some Mark V tanks crossed a muddy field without sinking the tracks over five inches, where we in the M4 started across the same field the same day and bogged down."

The 90 mm M36 tank destroyer was finally introduced in September 1944; the 90 mm round also proved to have difficulty penetrating the Panther's glacis plate, and it was not until an HVAP version of the round was developed that it could effectively penetrate it from combat range. It was very effective against the Panther's front turret and from the side, however.

The high U.S. tank losses in the Battle of the Bulge against a force composed largely of Panther tanks brought about a clamour for better armour and firepower. At General Eisenhower's request, only 76 mm gun-armed M4 Shermans were shipped to Europe for the remainder of the war. Small numbers of the M26 Pershing were also rushed into combat in late February 1945. A dramatic newsreel film was recorded by a U.S. Signal Corps cameraman of an M26 successfully stalking and then knocking out a Panther in the city of Cologne, however only after the Panther had already knocked out two M4 Shermans.

Production of Panther tanks and other German tanks dropped off sharply after January 1945, and eight of the Panther regiments still on the Western Front were transferred to the Eastern Front in February 1945. The result was that for the rest of the war during 1945, the greatest threats to the tanks of the Western Allies were no longer German tanks, but infantry anti-tank weapons such as the 88 mm calibre *Panzerschreck* (the German bazooka) and *Panzerfaust* anti-tank grenade launcher, and infantry anti-tank guns such as the ubiquitous 7.5 cm Pak 40, and mobile anti-tank guns such as the Marder, StuG III, StuG IV, and *Jagdpanzer*. A German Army status report dated March 15, 1945 showed 117 Panthers left in the entire Western Front, of which only 49 were operational.

According to US Army Ground Forces statistics, destruction of a single Panther was achieved after destruction of 5 M4 Shermans or some 9 T-34s.

DESIGN CHARACTERISTICS

Hitler personally reviewed the final designs for the Panther and it was he who insisted on an increase in the thickness of the frontal armour. Under his orders the front glacis plate was increased from 60 mm to 80 mm and the turret front plate was increased from 80mm to 100 mm.

As a result of Hitler's intervention the weight of the production model was increased to 45 metric tons an increased by 10 tons from the original plans for a 35 ton tank. To exacerbate matters the Panther was rushed into combat before all of its teething problems were corrected. Reliability was considerably improved over time, and the Panther did prove to be a very effective fighting vehicle; however, some design flaws, such as its weak final drive units were never corrected due to various shortages in German war production.

THE MAYBACH ENGINE

The first 250 Panthers were powered by a Maybach HL 210 P30 engine, a V-12 petrol engine which delivered 650 hp at 3,000 rpm and was protected by three simple air filters. Starting in May 1943, the next run of Panthers were built using the 700 PS (690 hp, 515 kW)/3000 rpm, 23.1 litre Maybach HL 230 P30 V-12 petrol engine. The designs of both engines were excellent and gave a remarkably high output for such a compact device. Two multistage "cyclone" air filters were used to automate some of the dust removal process. Once more however the increasingly difficult supply system encroached and the British control of aluminium supplies from Turkey dictated that the light alloy block used in the HL 210 was soon replaced by a less effective cast iron block. This was done to preserve the limited aluminium supply which was desperately needed elsewhere particularly in the production of jet engines. In practice the engine power output of the engines employed in the Panther was reduced due to the use of low grade petrol. With a full tank of fuel, a Panther could in theory cover 130 km on surfaced roads and 80 km cross country.

The HL 230 P30 engine was a very compact design, which kept the space between the cylinder walls to a minimum. The crankshaft comprised of seven discs, each with an outer race of roller bearings, and a connecting crankshaft pin between each disc. To reduce the length of the engine further, by one half a cylinder diameter, the usual practice was abandoned and the two banks of 6 cylinders of the V-12 were not offset. The centre points of the connecting rods of each cylinder pair in the "V" where they joined the crankshaft pin were thus at the same spot rather than offset; to accommodate this arrangement, one connecting rod in the pair of cylinders was forked and fitted around the other "solid" connecting rod at the crankshaft pin. (A more typical "V" engine would have had offset cylinder banks and each pair of connecting rods would have fit simply side by side on the crankshaft pin).

This unusual arrangement with the connecting rods was the source of considerable teething problems.

The cylinder head gaskets were another major problem and the combination of poor fuel and lubricants led to a large instance of blown head gaskets. This was one problem which could be corrected with the introduction of improved seals from September 1943. Another advance lay in the improved bearings which were introduced in November 1943. In common with the Tiger I it was soon discovered that allowing the engine speed to rise to 3000 rpm led to catastrophic failures. The obvious solution was to incorporate an engine governor which was added in November 1943. This essential device reduced the maximum engine speed to 2500 rpm. The situation was further improved by the addition of an eighth crankshaft bearing which was added to the production process beginning in January 1944. This too helped to reduce the previously high rate of motor engine failures.

The weight of the Panther posed major problems for bridge crossings. Like the Tiger I the engine compartment space of the

The importance of engine coolants was strongly emphasised in the Pantherfibel.

Panther was therefore designed to be watertight so that the Panther could be submerged and cross waterways. The consequence of this was that the engine compartment was poorly ventilated and prone to overheating. In addition the fuel connectors in the early models were non-insulated, leading to leakage of fuel fumes into the engine compartment. This unfortunate combination was the source of many engine fires which blighted the deployment of the early Panthers. The solution was to add additional ventilation venting through the engine deck which was designed to draw off these gasses. To an extent this reduced the instance of engine fires but it did not completely solve the problem and engine fires continued to claim precious Panther tanks. Other measures taken to reduce this problem included improving the coolant circulation inside the motor and adding a reinforced membrane spring to the fuel pump. As far as the crews were concerned it was fortunate that the Panther had a solid firewall separating the engine compartment and the fighting compartment in order to keep engine fires from spreading.

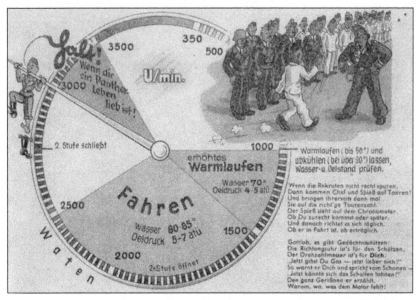

The importance of keeping the engine revs in a band between 1500 and 2500 is graphically demonstrated in the Pantherfibel.

The simple gearing for the final drive system of the Panther was easy to manufacture but was far less robust than the relatively complex gearing system on the Tiger.

The engines fitted into the Panther undoubtedly became more reliable over time, but as events demonstrated there was simply not enough time. In the aftermath of World War II a French assessment of their stock of captured Panthers conducted in 1947 concluded that the engine had an average life of 1,000 km and maximum life of 1,500 km.

SUSPENSION

The suspension system of the Panther closely resembled that of the Tiger I and consisted of two front drive sprockets, two rear idlers and eight double-interleaved rubber-rimmed steel road wheels on each side. The road wheels were suspended on a dual torsion bar suspension. The dual torsion bar system was designed by Professor Ernst Lehr and was purpose designed to allow for a wide travel stroke and rapid oscillations with high reliability.

The result of the innovative dual torsion bar system was meant that it was possible for the Panther to attain a relatively high cross country speed and the impressive ability to travel at high speed cross country was a defining feature of this remarkable heavy tank. The high speed of the Panther could be maintained over undulating terrain. However, the speed of the Panther came at a very high price. The extra space required for the bars running across the length of the bottom of the hull, below the turret basket significantly increased the overall height of the tank and also prevented the incorporation of an escape hatch in the hull bottom. When damaged by mines, the finely engineered torsion bars were easily bent out of shape required a welding torch for removal.

The Panther's suspension was complicated to manufacture and in common with the Tiger I incorporated the interleaved system which required the outer wheels to be removed in order to access the rear wheels and made replacing inner road wheels time consuming. The crews of the Panther would no doubt have been relieved to discover that the road wheels on their vehicle were

Taken in Northen France in October 1943, this photograph clearly shows the interleaved wheels of the Panther.

arranged in just two rows as opposed to the three of the Tiger I.

One tiresome feature of the interleaved wheels was that they exhibited a tendency to become clogged with mud, snow and ice, and could easily freeze solid overnight in the harsh winter weather of the Eastern Front. Shell damage could also cause the road wheels to jam together and become extremely difficult to separate. Interleaved wheels had long been standard on all German half-tracks. The extra wheels did provide better flotation and stability, and also provided more armour protection for the thin hull sides than smaller wheels or non-interleaved wheel systems, but the complexity and the tedious processes involved in maintenance meant that no other country ever adopted this cumbersome design for their tanks.

The road wheels of the Panther were rubber rimmed but in September 1944, and again in March/April 1945, M.A.N. the shortages of this vital substance led to the building of a limited number of Panther tanks with steel road wheels which were originally designed for the Tiger II and late series Tiger I tanks. Steel road wheels were introduced from chassis number 121052.

Once the Allied air forces began targeting Schweinfurt the resultant shortage of ball bearings was another major issue and in consequence, from November 1944 through February 1945, an emergency conversion process began which revolved around the use of sleeve bearings as an alternative to ball bearings. The sleeve bearings were primarily used in the running gear although contingency plans were made should the need arise to convert the transmission to sleeve bearings, but these were not carried out as production of Panther tanks came to an end.

STEERING AND TRANSMISSION

In the Panther, steering was accomplished through a seven-speed AK 7-200 synchromesh gearbox. It was designed by

Zahnradfabrik Friedrichshafen, and incorporated a MAN single radius steering system which, unlike the Tiger I with its steering wheel operation, the Panther utilised the traditional arrangement of steering levers.

On the Panther each gear had a fixed radius of turning, ranging from five meters for 1st gear up to 80 meters for 7th gear. The driver was expected to anticipate the sharpness of a turn and shift into the appropriate gear to turn the tank. The driver also had the option of engaging the brakes on one side to force a sharper turn. This manual steering was a much simplified design, compared to the more sophisticated dual-radius hydraulically controlled steering system of the Tiger I and ease of manufacturing compared to the Tiger I was therefore much enhanced. The AK 7-200 transmission was also capable of pivot turns, but this method of turning placed a great deal of additional strain which could accelerate failures of the final drive.

Throughout its career, the weakest part of the Panther was its final drive unit. The problems arose from a combination of

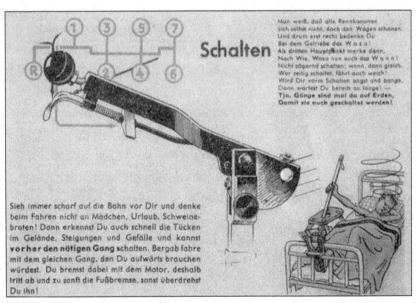

The seven forward and one reverse gear of the Panther from the Pantherfibel.

factors. The original MAN proposal had called for the Panther to have an epicyclic gearing (hollow spur) system in the final drive, similar to that used in the Tiger I. However, Germany at the time suffered from a shortage of gear-cutting machine tools and, unlike the Tiger tanks, the Panther was intended to be produced in large numbers. To achieve the goal of higher production rates, numerous simplifications were made to the design and its manufacture. This process was aggressively pushed forward, sometimes against the wishes of designers and army officers, by the Chief Director of Armament and War Production, Karl-Otto Saur (who worked under, and later succeeded, Reichminister Speer). Consequently, the final drive was changed to a more simple double spur system. Although much simpler to produce, the double spur gears had inherently higher internal impact and stress loads, making them prone to failure under the high torque requirements of the heavy Panther tank. Furthermore, high quality steel intended for double spur system was not available for mass production, and was replaced by 37MnSi5 tempered steel, which was unsuitable for such a high stress gearing arrangement. In contrast, both the Tiger II and the US M4 Sherman tank had double helical (herringbone gears) in their final drives, a system that reduced internal stress loads and was less complex than epicyclic gears.

Compounding these problems was the fact that the final drive's housing and gear mountings were too weak because of the poor type of steel available and the tight space allotted for the final drive. The final gear mountings deformed easily under the high torque and stress loads, pushing the gears out of alignment and resulting in failure. Due to the weakness of the final drives their average fatigue life was only 150 km. In Normandy, about half of the abandoned Panthers were found by the French to have broken final drives. However, at least the final gear housing was eventually replaced with stronger one, while final gear problem was never solved.

Plans were made to replace the final drive, either with a version of the original epicyclic gears planned by MAN, or with the final drive of the Tiger II. These plans were intertwined with the planning for the Panther II, which never came to fruition because Panzer Commission deemed that temporary drop in production of Panther due to merger of Tiger II and Panther II was unacceptable. It was estimated that building the epicyclic gear final drive would have required 2.2 times more machining work than double spur gears, and this would have affected manufacturing output.

Most of the shortcomings were considered acceptable once design flaws were rectified. Due to the mechanical unreliability of final gear the Panther had to be driven by experienced drivers with extreme care, a characteristic shared with the Tiger tanks as well as *Jagdtigers*. Long road marches would inevitably result in a significant number of losses due to breakdowns, and so the German Army had to ship the tanks by rail as close to the battlefield as possible. This theoreticaly convenient and sensible arrangement was not always achievable in practice and the

The turning radius of each of the Panther's forward gears from the Pantherfibel.

Repair of the transmission of a Panther, Russia, May 1944.

Panthers continued to face unfeasibly long road marches which led to numerous breakdowns.

THE PANTHER II

The early impetus for upgrading the Panther came from the concern of Hitler and others that it lacked sufficient armour. Hitler had already insisted on an increase in its armour once so far and further discussions involving Hitler, in January 1943, resulted in a call for further increased armour; initially referred to by an Arabic numeral as the Panther 2, it was redesignated with the Roman numeral becoming the Panther II after April 1943. This upgrade increased the glacis plate to 100 mm, the side armour to 60 mm , and the top armour to 30 mm. Production of the Panther 2 was slated to begin in September 1943.

In a meeting on February 10, 1943, further design changes were proposed - including changes to the steering gears and final drives. Another meeting on February 17, 1943 focused on sharing and standardizing parts between the Tiger II tank and the Panther 2, such as the transmission, all-steel roadwheels, and running gear. Additional meetings in February began to outline the various components, including use of the 88 mm L/71 KwK 43 gun. In March 1943, MAN indicated that the first prototype would be completed by August 1943. A number of engines were under consideration, among them the new Maybach HL 234 fuel-injected engine (900 hp operated by an 8-speed hydraulic transmission).

It was a sign of the rapid pace of events that the up-grade path to replace the original Panther design with the Panther II were already underway even before the first Panther had even seen combat. However from May to June 1943, work on the Panther II ceased as the focus shifted to expanding production of the original Panther tank. It is not clear if there was ever an official cancellation - this may have been because the Panther II upgrade pathway was originally started at Hitler's insistence. The direction that the design was headed would not have been consistent with Germany's need for a mass-produced tank, which was the goal of the Reich Ministry of Armament and War Production.

One Panther II chassis was completed and eventually captured by the U.S.; it is now on display at the Patton Museum in Fort Knox. An Ausf. G turret is mounted on this chassis.

PANTHER AUSF. F

After the Panther II project was abandoned, a more limited upgrade of the Panther was planned, centered around a re-designed turret. The Ausf. F variant was slated for production in April 1945, but the war ended these plans.

The earliest known redesign of the turret was dated November 7, 1943 and featured a narrow gun mantlet behind a 120 mm

A knocked out Panther tank lies redundant in the river at Houffalize, 1945.

(4.7 in) thick turret front plate. Another design drawing by Rheinmettall dated March 1, 1944 reduced the width of the turret front even further; this was the *Turm-Panther* (*Schmale Blende*) (Panther with narrow gun mantlet). Several experimental *Schmalturm* (literally: "narrow turret") were built in 1944 with modified versions of the 75 mm KwK 42 L/70, which were given the designation of KwK 44/1. A few were captured and shipped back to the U.S. and Britain. One is on display at the Bovington tank Museum.

The *Schmalturm* had a much narrower front face of 120 mm (4.7 in) armour sloped at 20 degrees; side turret armour was increased to 60 mm (2.4 in) from 45 mm (1.8 in); roof turret armour increased to 40 mm (1.6 in) from 16 mm (0.63 in); and a bell shaped gun mantlet similar to that of the Tiger II was used. This increased armour protection also had a slight weight saving due to the overall smaller size of the turret.

The Panther Ausf. F would have had the *Schmalturm*, with its better ballistic protection, and an extended front hull roof which was slightly thicker. The Ausf. F's *Schmalturm* was to have a built-in stereoscopic rangefinder and lower weight than the original turrets. A number of Ausf. F hulls were built at Daimler-Benz and Ruhrstahl-Hattingen steelworks; however there is no evidence that any completed Ausf. F saw service before the end of the war.

Proposals to equip the *Schmalturm* with the 88mm KwK 43 L/71 were made from January through March 1945. These would have likely equipped future German tanks but none were built, as the war ended.

ABOUT CODA BOOKS

Most Coda books are edited and endorsed by Emmy Award winning film maker and military historian Bob Carruthers, producer of Discovery Channel's Line of Fire and Weapons of War and BBC's Both Sides of the Line. Long experience and strong editorial control gives the military history enthusiast the ability to buy with confidence.

The series advisor is David McWhinnie, producer of the acclaimed Battlefield series for Discovery Channel. David and Bob have co-produced books and films with a wide variety of the UK's leading historians including Professor John Erickson and Dr David Chandler.

Where possible the books draw on rare primary sources to give the military enthusiast new insights into a fascinating subject.

www.codabooks.com

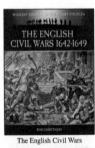

The English Civil Wars

The Zulu Wars

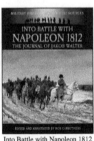

Into Battle with Napoleon 1812

Waterloo 1815

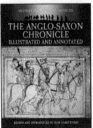

The Anglo-Saxon Chronicle

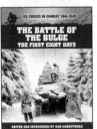

The Battle of the Bulge

The Normandy Campaign 1944

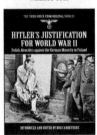

Hitler's Justification for WWII